GOD and EVOLUTION

JÓZEF ŻYCIŃSKI

GOD *and* EVOLUTION
Fundamental Questions of Christian Evolutionism

Translated by KENNETH W. KEMP
and ZUZANNA MAŚLANKA

The Catholic University of America Press
Washington, D.C.

Originally published as *Bóg i ewolucja: Podstawowe pytania ewolucjonizmu chrześcijańskiego*. Lublin, Poland: Towarzystwo Naukowe Katolickiego Uniwersytetu Lubelskiego, 2002.

Copyright © 2006
The Catholic University of America Press
All rights reserved
The paper used in this publication meets the minimum requirements of American National Standards for Information Science—Permanence of Paper for Printed Library Materials, ANSI Z39.48-1984.
∞

LIBRARY OF CONGRESS CATALOGING-IN-PUBLICATION DATA
Zycinski, Józef.
 [Bóg i ewolucja. English]
 God and evolution : fundamental questions of Christian evolutionism / Józef Zycinski ; translated by Kenneth W. Kemp and Zuzanna Maslanka.
   p. cm.
 Includes bibliographical references.
 ISBN-13: 978-0-8132-1470-2 (pbk. : alk. paper)
   ISBN-10: 0-8132-1470-X (pbk. : alk. paper)   1. Evolution—Religious aspects—Christianity.   I. Title.
 BL263.Z9313 2006
 231.7'652—dc22
                                                                    2006006975

CONTENTS

Introduction: *Interdisciplinary Dialogue in Place of Pathology*   1

PART ONE

1. Biology and Metaphysics in Charles Darwin's Conception of Evolution   11
   The Sphere of Inductive Epistemology   11
   The Co-existence of Science and Faith   14
   Theological Motifs in the Thought of Darwin   17
   The Search for a Theology of Darwinism   25

2. Fundamentalisms and Evolution   32
   The Origin of Christian Fundamentalism   32
   Fundamentalist Interpretation of Biblical Texts   36
   The Fundamentalism of Phillip E. Johnson   39
   Fundamentalism and Catholicism   42

3. Elements of Fundamentalism in Atheistic Evolutionism   45
   Elements of Agnosticism and Atheism in Evolutionary Views of Nature   47
   Religious Agnosticism and the Principle of Ockham's Razor   51
   Epistemological Clarity in Place of Fundamentalisms   57

4. Evolution and Christian Thought in Dialogue according to the Teaching of John Paul II   60
   Evolutionism according to the Message to the Pontifical Academy of Sciences   61
   Philosophical Questions of Evolutionism   64
   Scientific Emergentism and Ontological Emergentism   66
   The Immanent Divine Logos   70

GOD *and* EVOLUTION

Introduction

INTERDISCIPLINARY DIALOGUE IN
PLACE OF PATHOLOGY

[Science and faith] are two distinct autonomous trajectories, but by their very nature they are never on a collision course. Whenever some type of friction is noted, it is a symptom of an unfortunate pathological condition.[1]

JOHN PAUL II

Over the course of time, the question of the harmonious unification of scientific theories of anthropogenesis with Christian faith in a Creator who directs the processes of evolution has received new and more insightful answers. This development is possible thanks to new discoveries concerning both cosmic and biological evolution. The introduction of new concepts and subtle distinctions allows one to avoid those oversimplified contradictions of the past, in which God was supposed to act on nature only through extraordinary interventions while the application of deterministic explanations was supposed to exclude definitively the possibility of appeal to any kind of teleological categories. To those changes one must add the desire, characteristic of the pontificate of John Paul II, for interdisciplinary

---

1. "Faith and Science: Gift of God," Address of Pope John Paul II to the international scientific community during a visit to Ettore Maiorana Research Centre (8 May 1993), published in *The Pope Speaks* 39 (1993): 5: 297.

dialogue among the natural sciences, philosophy, and theology. A document especially important for this topic is the letter of the Holy Father to the director of the Vatican Astronomical Observatory, George Coyne, S.J.[2] Ernan McMullin, the well-known Notre Dame philosopher of science, calls the document "the most important Roman statement on this topic since Pius XII's address to the Pontifical Academy of Sciences in 1951."[3] In biological circles, the Papal message on the theory of evolution, addressed to the Pontifical Academy of Sciences on 22 October 1996, evoked even stronger reactions. Many prestigious periodicals in theoretical biology gave their attention to that message, emphasizing that the teachings of John Paul II cannot in any way be brought into agreement with the position of creation science or with various versions of contemporary Biblical fundamentalism which defend the literal interpretation of Holy Scripture.[4]

In itself the rejection of fundamentalisms does not lead to the elimination of all controversies associated with the theory of evolution. The quarrel between methodological naturalism and ontological naturalism appears as a fundamental line of division. The former is a necessary condition for conducting science according to the methodology worked out in the period of Galileo and Sir Isaac Newton. The latter expresses a powerful metaphysics, which its proponents attempt to introduce under the guise of new scientific theories showing the mechanisms of evolution. That metaphysics is the expression of

---

2. John Paul II, "Message to the Reverend George V. Coyne, S.J., Director of the Vatican Observatory, June 1, 1988," in R. Russell, W. Stoeger, and G. Coyne, eds., *Physics, Philosophy and Theology: A Common Quest for Understanding* (Vatican City: Vatican Observatory, 1988), M7 and M8.

3. E. McMullin, "A Common Quest for Understanding," in R. J. Russell, W. R. Stoeger, and G. V. Coyne, eds., *John Paul II on Science and Religion: Reflections on the New View from Rome* (Vatican City: Vatican Observatory, 1990), 53.

4. *The Quarterly Review of Biology* devotes an entire issue—number 72 (1997)—to the Papal evaluation of the theory of evolution as well as to commentary on it from the pens of Edmund Pelegrino, Michael Ruse, and Richard Dawkins. Only Dawkins is generally critical in his remarks. On the other hand, Ruse concludes his evaluation with the statement: "Were I a Catholic, I would positively welcome Darwin as an ally" (p. 394).

In our evaluation of such conclusions, we must not forget that the human brain and its culture-creating potential was also formed as a result of biological evolution. Therefore, in the name of objectivity, we must not introduce any general axiom about the superiority of cultural evolution over biological in the total process of the evolution of nature, but we must at most limit ourselves to a more careful thesis about the particular role of cultural factors at the level of the evolution of man. Such a thesis would make a contribution to our understanding of the evolutionary peculiarities of the species *Homo sapiens*.

One must also remember that not only the speed of the development of thought, but also its content and its significance, are important to the future of our species. There are those who have tried to reduce the content of our culture to mechanisms in genes' struggle for existence. Today, when the radical and unverifiable slogans of the sociobiologists have fallen into oblivion—and with the rejection of all versions of fundamentalism as interpretive disorders—special conditions exist for concentrating attention on the question: What role can contemporary *Homo sapiens* play in building a culture which expresses man's special position in an evolving nature?

PART ONE

German liberals, science was being used in the development of social utopias with a marked antireligious cast, in English scientific institutions every effort was made to gain acceptance of the genuine discoveries of the new sciences among the working classes. In a series of public lectures having as their goal the popularization of the new scientific discoveries, James Ferguson was introducing his fascinated female audience to the mysteries of Newtonian mechanics, while Adam Sedgwick was telling mechanics about the habits of South American lizards. The demand for such popularizations of knowledge was so great that Lord Arthur Hervey, Bishop of Bath and Wells, encouraged universities to create special chairs whose occupants would concern themselves exclusively with popular educational activity. The ambitious program of Bishop Hervey was inspired by his conviction that the purpose of academic institutions was not only to form the intellectual elite but also to influence popular consciousness. It was hoped that, among the important results of such influence would be formation of the conviction that "a knowledge of nature goes hand in hand with a knowledge of the word of God," "Christian truth has the full approval of an enlightened reason," and "the truths of science attain a fresh gleam of beauty" thanks to their directedness towards Divine Wisdom.

The position taken by Archbishop Frederick Temple in his Bampton Lectures must be recognized as authoritative in the atmosphere of English academic institutions of that period. Delivering a series of lectures at Oxford in 1884 on the theme *The Relations between Religion and Science,* the future Archbishop of Canterbury presented the evolutionary view of nature as a scientific certainty about which any discussion would be useless and without foundation.[5] The categori-

---

5. Cf. O. Chadwick, "Evolution and the Churches," in C. A. Russell, ed., *Science and Religious Belief: A Selection of Recent Historical Studies* (London: University of London Press, 1973), 282.

curred having been preordained for a special end, I can no more believe in it than that the spot on which each drop of rain falls has been specially ordained.⁹

This confession stands in contrast to the reminiscences of his student days in Cambridge which he recorded in his *Autobiography*. Darwin there became familiar with the works of William Paley, in which the classical formulation of the argument from design found expression. Many years later, he would admit that the clear logic of Paley's conclusions from *Natural Theology* gave him the feeling of elation which he had earlier experienced in leafing through the pages of Euclid's *Elements;* he was both charmed and convinced by this text.

In the writings of the later Darwin, these same arguments met with an unequivocal skepticism. What dominates in these writings is a questioning of traditional pictures of that order of the world in which design is connected with immediate interventions of the Creator. However, one can also see there a rejection of the idea that the order of universal laws of nature is a result of the play of accidental conditions. On the contrary, Darwin was inclined to the opinion that the observed order of nature expresses an order defined by God in a general way, and not in a concrete way dependent on particular physical conditions. Experiencing deep philosophical perplexities, Darwin wrote:

I cannot anyhow be contented to view this wonderful universe, and especially the nature of man, and to conclude that everything is the result of brute force. I am inclined to look at everything as resulting from designed laws, with the details, whether good or bad, left to the working out of what we may call chance. Not that this notion at all satisfies me.[10]

---

9. Letter of 12 July 1870 to J. D. Hooker, in C. Darwin, *More Letters of Charles Darwin*, F. Darwin and A. C. Seward, eds. (New York: Appleton, 1903), 1: 321.
10. Letter of 22 May 1860 to Asa Gray, in *Correspondence* 8: 224.